Lorena Dutra Silva

Analysis of the chemical profile of the leaves and essential oil of Eugenia unifl

Lorena Dutra Silva

Analysis of the chemical profile of the leaves and essential oil of Eugenia unifl

And in silico modelling of the kinetic profile

ScienciaScripts

Imprint
Any brand names and product names mentioned in this book are subject to trademark, brand or patent protection and are trademarks or registered trademarks of their respective holders. The use of brand names, product names, common names, trade names, product descriptions etc. even without a particular marking in this work is in no way to be construed to mean that such names may be regarded as unrestricted in respect of trademark and brand protection legislation and could thus be used by anyone.

Cover image: www.ingimage.com

This book is a translation from the original published under ISBN 978-613-9-63371-5.

Publisher:
Sciencia Scripts
is a trademark of
Dodo Books Indian Ocean Ltd. and OmniScriptum S.R.L publishing group

120 High Road, East Finchley, London, N2 9ED, United Kingdom
Str. Armeneasca 28/1, office 1, Chisinau MD-2012, Republic of Moldova, Europe
Printed at: see last page
ISBN: 978-620-7-66665-2

EPIGRAPH

"Maybe I didn't manage to do the best, but I fought for the best. I'm not what I should be, but thank God I'm not what I was before." Marthin Luther King

SUMMARY

Eugenia uniflora is described as having therapeutic properties that help to reduce the risk of various chronic degenerative diseases such as cardiovascular disease and cancer. This effect has been attributed to the presence of antioxidant compounds, such as phenolic compounds, vitamin C, carotenoids, and secondary metabolites from plants, called polyphenols. The aim of this study was therefore to assess the chemical composition of the leaves and essential oil of Eugenia uniflora, as well as to evaluate its metabolism and pharmacokinetic parameters in silico. To this end, after collection in February and September 2017, a phytochemical profile was carried out according to the methodology of Wagner & Bladt, (1996) and Abreu (2000) and the essential oil was analysed by GC-MS. The phytochemical analysis revealed the presence of secondary metabolites such as cyanidin, phenolic heterosides and coumarins. The essential oil analysis revealed 20 compounds in February and 10 compounds in September. Abiotic factors were observed to interfere with the results in terms of chemodiversity and quantity. There were five main compounds in February and three in September: curzerene 1(31.39%; RT: 28.29) 2(78.12%, RT:27.836); Y- elemene 1(21.21%; RT: 25.635) 2(7.56%, RT:30.217), P- germacrene 1(17.52%; RT:30.786) 2(5.90%, RT:35.363); germacrone (5.79%; RT: 35.829), õ-germacrene (5.09%; RT: 27.688). The *lipinks* violation rule was used to see if the metabolites were considered possible good drugs, and only metabolite 3 of curzerene had two violations. Thus, it was observed that Eugenia uniflora presented a chemodiversity of 20 chemical compounds in January and 10 compounds in August, with the main active phytochemical compounds being coumarin, digitalis and saponins. They therefore presented interesting secondary metabolites with great biological potential.

Keywords: medicinal plants, phytochemistry and pharmacokinetics

SUMMARY

INTRODUCTION

The promising use of plants in the prevention and treatment of various pathologies can be explained by the diversity of organic compounds and phytoconstituents they contain, which can have similar or even superior efficacy to synthetic compounds (CECHINEL FILHO et al. 2003). For this reason, this therapeutic alternative has traditionally been used to treat various diseases, including infectious processes and cancer (CALIXTO, 2003).

In Brazil, even with the constant encouragement of the media and the manufacturers of

While many people are still using complementary practices to look after their health, such as the use of medicinal plants, which are used for prevention, as a palliative and to cure some illnesses. Currently, economic, political and social changes have influenced not only people's health but also the model of care. In order to cater for this new model of care and guarantee the Brazilian population safe access to and rational use of medicinal plants, the National Policy for Integrative and Complementary Practices (PNPIC) was instituted in the SUS by the Ministry of Health (MS) (WHO, 1993; 2011).

Eugenia uniflora Lineu is popularly known as pitangueira, belongs to the Myrtacea family, and has been popularly used for its anti-hypertensive effect, vaso-relaxation of the thoracic aortic rings, diuretic activity and cardiovascular activity (CONSOLINI & SARUBBIO, 2002; WAZLAWIK et al., 1997).

As such, this proposal will contribute scientifically to expanding data on the chemical composition and analysing the pharmacokinetic profile of the compounds elucidated from the essential oil of Eugenia uniflora leaves. Another aspect that demonstrates the scientific contribution of this proposal is that knowledge of plant composition and kinetics in the human body are essential requirements for transforming medicinal plants into phytotherapeutic products. Thus, plant research has been and continues to be an important alternative in the search for new drugs with therapeutic properties. Therefore, the expansion of scientific production on medicinal plants is of great importance, as this will create a scientific basis for the production and prescription of plant drugs in which efficacy and toxicity can be predicted (CECHINEL FILHO; YUNES, 1998).

CHAPTER 1

THEORETICAL FRAMEWORK

1.1 Medicinal plants

Various civilisations have left evidence or reports of the use of plant species with therapeutic potential. These plants played an unquestionable role in the prevention and treatment of the most varied diseases (VIOLANTE, 2008). This has been and continues to be possible due to the perpetuation and distribution of species, the succession of popular knowledge from one generation to the next, the advance of science and its collaboration in the knowledge of medicinal properties, as well as the development of tools that make it possible to standardise the use of medicinal plants in therapy (RIEDER; GUARIM NETO, 2012).

Currently, many factors have contributed to the increased use of plants in therapy, including the high cost of synthetic allopathic medicines, the population's difficult access to medical care, the increased resistance of tumours and microorganisms to traditional chemotherapy, the emergence of new pathologies and the need for palliative treatment of known pathologies without curative treatment in the medical field through the use of plants (YUNES; CECHINEL FILHO, 2014; FELICIANO; CECHINEL FILHO, 2014).

Medicinal plants have always been the object of study by scientists looking for new sources to obtain the active ingredients responsible for their pharmacological or therapeutic action on the population, and a cheaper way of curing and preventing diseases (HAMILTON, 2004;

LORENZI, 2008; MATOS, 2008). Various parts of plants can be used in medicinal treatment: leaves, fruit, bark, seeds, roots and flowers. These plants can be used as they are or in the form of aqueous extracts, alcoholic extracts, plasters and others. They can be produced using techniques such as: catoplasm, decoction, inhalation, infusion, potions and so on (SOSSAE, CRISTINA F., 2008).

Most plants with medicinal properties are easy for communities to grow and access. Many have widespread popular use, however, in most cases there is still a lack of studies proving the medicinal properties, efficacy dose and toxicity of these species, which is necessary for effective treatment, rational use and patient safety (PLANTAS MEDICINAIS, 2009).

In this context, Brazil is described as a country with great potential in terms of natural resources with medicinal properties. Among the various biomes in Brazil, the one with the greatest diversity of medicinal plants is the Amazon rainforest, which is home to 50 per cent of the planet's biodiversity. Surpassed only by the Amazon Rainforest, the Cerrado is Brazil's second largest biome in terms of extension. It is possible to find at least 200 native plants with economic or medicinal potential, mainly species with anti-tumour, anti-fungal and anti-inflammatory activity (IPEF, 2006). Most of these medicinal properties are used by local communities and have great economic potential. Some of the most widely used are from the Myrtacea family, such as the fruit trees: jaboticaba (Eugenia cauliflora

O. Berg), guava trees (Psidium guajava L.). Aromatic trees such as: Indian carnation (Syzygium aromaticum L. Merr. Et Perry). Those that offer quality wood used in construction, such as: jequitibá-rosa (Couratari legalis Mart.), jequitibá-vermelho (C. estrellensis Ames) among others (PINTO, 1956; MAEDA et al. 1990). It is one of the most complex families from a taxonomic point of view, both in terms of the number of species and also due to the scarcity of studies (SOUSA; LORENZI, 2008).

1.2 . Myrtaceae

Myrtaceae is a botanical family comprising 130 genera and around 3,000 species. They are shrubs or trees represented in the Americas mainly by fruit-bearing plants such as jambo, pitanga, uvalha, guava, araçá and jabuticaba (MYRTACEAE GENERA, 2006). All the species in this family are woody and generally vary in size from 2 metres in height (shrubs) to several tens of metres in height (mega phanerophytes), with tissues rich in essential oils and flowers with a multiple of 4 or 5 floral pieces. The leaves are simple, opposite, entire-edged, pinnate and generally with a marginal vein. In Brazil it is the richest of the angiosperm families and the largest contributor to the Brazilian flora. Found mainly in the easily accessible Cerrado biome in Goiás, it includes 20 genera and 202 species, many new species continue to be described every year from the vast regions where the Myrtaceae are found. For the same reasons, new genera are described almost every year and the taxonomic perimeters of others remain unfixed, leading to frequent revisions (ANGIOSPERM PHYLOGENY GROUP III, 2009).

Some of the main genera in the Myrtaceae family are Eugenia L. Eucalyptus L. and Psidium L. and Myrcia DC. As shown in Table 1 below, the Eugenia L. genus has the largest number of species and is the most abundant in the Cerrado biome (SOUSA; LORENZI, 2008).

Table 1: Genera of the Myrtaceae family.

Gender	Species	Medicinal properties
Eugenia	*E. uniflora L.* *E. dysenterica DC.* *E. jambolana*	Potential to destroy cancer cells, control diabetes and cholesterol.
Eucalyptus	*Eucalyptus angophora L,* *Eucalyptus corymbia L,* *Eucalyptus globulus L.*	Treatment of flu, nasal congestion and sinusitis.
Psidium	*Psidium guajava L.* *Psidium myrsinoides* 0. fíerg *Psidium pohlianum Berg*	Natural antiseptic, diarrhoea, dysentery and Hypoallergenic
Myrcia	*Myrcia tomentosa DC.* *Myrcia multiflora DC.* *Mycia dúbia DC.*	Diabetes control, diarrhoea, canker sores, haemorrhoids, rich in vitamin C

Source: (LORENZI et al., 2006); (LORENZI; MATOS, 2002); (Araujo, G., 2017).

1.3 Genus Eugenia Lineu.

The genus Eugenia L. makes a great socio-economic contribution as it has a wide range of fruit species, leading to their commercialisation.

It is also rich in essential oils from which creams and perfumes are made, and its wood is used mainly for the eucalyptus trees, which are the most commercialised,

widely used as ornamental plants (FONTENELLE; COSTA; MACHADO, 1994).

It is widely used in folk medicine for its anti-inflammatory and antimicrobial activity in the treatment of communicable diseases (VELOSO; JUVENAL HENRIQUE, 2016). Some of the species in question are Eugenia uniflora, Eugenia dysenterica, Eugenia jambolana, Eugenia luschnathiana, Eugenia uvalha and Eugenia aggregata.

Found in the Cerrado, they have also been described as having antineoplastic and antioxidant potential, as shown in Chart 2 (MARON; STASI; MACHADO, 2006).

Table 2: Eugenias species with medicinal properties.

Scientific name	Popular name	Medicinal properties	Chemical compounds
E. jambolana	Jambolan	Potential to lyse cancer cells, fight gout and allergies.	Bioactive compounds against free radicals and lipid oxidation
E. dysenterica	Cagaita	Rich in vitamins C, antioxidants, laxative, purifying and astringent	Quercetin derivatives
E. uniflora	Pitanga	Fights premature ageing of cells	Presence of flavonoids

		and also cases of cancer	
E. luschnathiana	Pitomba	High vitamin C content	High antioxidant compounds
E. uvalha	Grapefruit	Vitamin A	Antioxidant compounds
E. *aggregata*	Rio Grande cherry	Diuretic, treatment of rheumatic and gastrointestinal diseases	Abundant in total anthocyanin content

Source: (SCHAPOVAL et al., 1994); (BRANDÃO, 1991); (ALBERTON et al.,2001).

1.4Eugenia		Uniflora

The Eugenia uniflora, popularly known as the pitangueira, belongs to

A member of the Myrtacea family, it has edible fruits that are much appreciated in Brazil and are used in juices, jellies and ice creams, and the tea made from its leaves is used in folk medicine, due to its reputation for having phenolic compounds with antioxidant action, as well as being hypotensive, antigout, stomachic and hypoglycaemic (ALMEIDA et al.2, 1995). Nowadays it is also widely used for its anti-hypertensive effect (CONSOLINI; SARUBBIO, 2002), vaso-relaxation of the thoracic aortic rings (WAZLAWIK et al., 1997), diuretic activity (CONSOLINI et al., 1999), cardiovascular activity (LEE et al., 2000) as well as anti-inflammatory, antimicrobial, cytotoxic and larvicidal activities (GARMUS et al., 2014).

As well as the leaves of Eugenia uniflora having beneficial chemical compounds, the fruit also has a wide range of compounds, such as caratenoids and phenolics, which are compounds that have beneficial effects on human health. There are also anthocyanins, flavonols and total carotenoids, which are promising sources of chemical compounds (LIMA, MELO, LIMA, 2002; AMORIM, 2009).

There are a variety of phytochemical compounds already identified in the leaves of the pitangueira species, such as flavonoids, anthraquinones, terpenes and tannins (FIUZA et al., 2008).

This family is rich in oil production. With a high incidence of

flavonoids such as myricetin and quercetin, which has led to much interest in studying this species (SCHMEDA-HIRSCHMANN, 1995). Maia et al. (1999) studied the essential oil of the leaves and branches of E. uniflora, harvested in the city of Belém, Pará, and obtained 1.8% essential oil, of which germacrene (32.8%), germacrene B (15.6%) and curzerene (30.0%) were the isolated compounds with the highest percentage content.

Melo et al. (2007) analysed the probable components of the essential oil of E. uniflora leaves responsible for its characteristic aroma. Using gas chromatography coupled with mass spectrometry, they obtained more than thirty compounds, including furanodiene and its rearrangement product, furnaoelemene, - elemene and a- cadinol, representing 50.2%, 5.9% and 4.7% respectively.

The essential oil of Eugenia uniflora is rich in sesquiterpenes such as germacrone, furodiene and curzerene, among others. It has attracted great attention from the scientific community for its compounds and in studies of its anti-cancer effects. The various pharmacological actions have yet to be properly investigated and proven, creating a gap that can and should be explored. (ZHONG, Z. 2001)

Figure 1: Trunk (A), foliage (B) and fruit (C) of E. uniflora. **Source**: Author, 2017.

1.5Importance of   chemodiversity

According to estimates by the Convention on Biological Diversity (CBD), Brazil is responsible for between 15 and 20 per cent of all the world's biodiversity, and is also considered the largest on the planet in terms of number of species. Brazil's biodiversity is considered a source of active biological substances (LOMBARDINO, J. G., 2004).

Preservation is essential both because of the immense biological wealth and also because it is a huge source of new drugs, which has aroused interest in various areas and countries because of this wealth. (LEWINSONHN, T. M., 2002.).

Since before Christ, there are records of natural medicine being used to treat the ailments of the time. Poppy, marijuana and aloe were used as the only resources available, which already had beneficial effects on human health. But it wasn't until the 19th century that research began into the active ingredients found in plants, which led to the creation of the first plant-based medicines that we now have at our disposal in the pharmaceutical industry (CALIXTO & SIQUEIRA JÚNIOR, 2008).

In 1803, the pharmacist Friedrich Wilhelm Adam Serturner marked the beginning of the extraction of the plant drug by managing to isolate morphine from Papaver somniferum. From these studies to the present day, various tests have been carried out to treat cancer, such as etoposide, vinblastine and vincristine, the latter two being alkaloids found in Catharanthus roseus (BULANAS and KINGHORN, 2005; BALICK and COX, 1996).

A milestone in the development of the current generation of drugs produced is the evolution of pharmaceutical companies that have been seeking new studies in search of new compounds, benefiting from the biotechnology technologies found today, favouring studies in chemistry and molecular biology and the application of recombinant DNA among many other areas, using engineering techniques, genetics, generating biotechnological drugs. (PALMEIRA FILHO & PAN, 2003).

It is believed that strains of microorganisms develop resistance when they are no longer destroyed or inhibited by the antibiotic and antiviral drugs that would usually be effective against them. In the same way, cancer cells can develop resistance to chemotherapy drugs. Each organism is different, so each individual reacts in some way, either by altering the structure of the cell or the biochemical pathways in a harmful way, altering the parts of the cell affected by the drugs, reducing their ability to act. This results in resistance (HUSSA A., 2017).

The discovery of new drugs of plant origin has encouraged research in this area, as it is producing great results in areas that were previously not given much importance. With drug resistance and outbreaks of new diseases, plant drugs are highly valued in this area of chemical studies, which has been developing the most in the scientific field. The species from the Amazon and the Cerrado have produced the most positive results. Especially those from the Cerrado (MENDES A. 2006).

1.   6Pharmacokinetic properties-in silico metabolism

The association of research into pharmacodynamics with data

Pharmacokinetic analyses offer a number of advantages. These advantages include understanding why unexpected effects occur, determining the ideal dose for a patient, anticipating data on the metabolisation and transport routes of substances similar to those studied and predicting events that may occur (TOZER; ROWLAND, 2009).

Four fundamental processes make up the body's pharmacokinetic profile: absorption, distribution, biotransformation and excretion (RAMOS; SILVEIRA, 2001). By understanding these events, it is possible to establish the chronological relationship between the pharmaceutical form, dose, frequency, route of administration and the systemic concentration of the drug, and clinical pharmacokinetics involves applying this data to patient therapy (TOZER & ROWLAND, 2009).

According to the International Union of Pure and Applied Chemistry (IUPAC), medicinal chemistry is responsible for inventing, planning, discovering and identifying biologically active compounds. By law, the safety and efficacy of drugs, even after highly successful compounds have been discovered, must be tested in in vitro or in vivo studies to characterise their biological effects (GUIDO et al., 2010).

Drugs are most often xenobiotics, which are used to modulate bodily functions for therapeutic purposes. Drugs and other environmental chemicals that enter the body are modified by a huge variety of enzymes present in our metabolism. These transformations can be altered by enzymes, making them beneficial, harmful, inactive or even toxic to the human body, which is why new antibiotics have been significantly decreasing on the market in recent years (KAPLAN, WIRTZ et al., 2013).

In silico is an expression used in computer simulation and related areas to indicate something that has happened. The expression originated from the Latin expressions in vivo and in vitro, often used in biology. In silico is originally used only to denote computer simulations that model a natural or laboratory process and not for generic computer calculations. In silico studies have the advantage of being quick to carry out, low cost and the ability to reduce the use of animals in toxicity tests (Danchin A, 1991).

These frequent biotransformations are called reaction I reaction II in two ways. Phase I enzymes take place mainly in the liver and make it possible to introduce functional groups, adding to xenobiotics such as: -OH, -COOH, -SH, -O- or NH2. The addition of these functional groups slightly increases the compound's water-solubility, but can profoundly alter its pharmacological properties. These reactions are strongly associated with the inactivation of active drugs. Phase II enzymes, on the other hand, are anabolic and involve conjugation with other components, as they facilitate the elimination of drugs and the inactivation of potentially toxic electrophilic metabolites produced by oxidation. These reactions produce more water-soluble metabolites with higher molecular weights, which facilitates their elimination (Rang, H P, et al.).

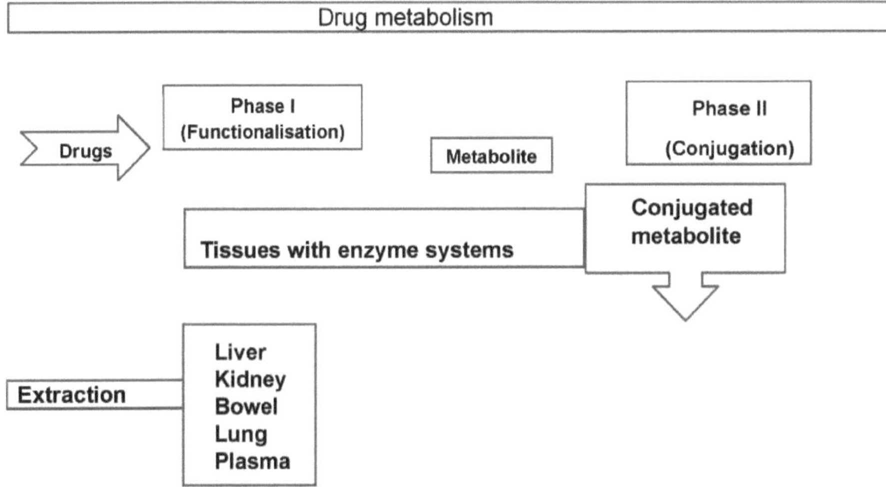

Figure 2: schematic demonstrating drug metabolism source: author

CHAPTER 2

OBJECTIVES

2 1 General

To identify the composition of the essential oil of Eugenia uniflora L. leaves and to evaluate in silico the pharmacokinetic properties of the essential oil's main compounds.

2.2 Specific

- To evaluate the phytochemical profile of *Eugenia uniflora L.* leaves and to extract essential oil from *Eugenia uniflora L.* leaves at two different times of the year.
- Analyse the chemical composition of the essential oil using gas chromatography coupled with mass spectrometry
- Evaluate the physical and chemical properties of the main compounds and their metabolism *in silico.*

CHAPTER 3

METHODOLOGY

3.1 Study Area

The municipality of Itapuranga is located in the state of Goiás in the Centre-West region of Brazil. Its territory is bordered by the municipalities of Heitoraí, Goiás, Guaraíta, Faina, Morro Agudo de Goiás, Uruana and Carmo do Rio Verde. It has an area of 1,277.15 km2 and an altitude of 600 metres above sea level (COSTA-JUNIOR; FIGUEIREDO, 2003).

The municipality's vegetation is made up of savannahs, fields and forests,

It is characterised by the following features: grasses that form the natural pastures, crooked and thick trees that spread out in a very irregular manner (GONÇALO; SOUZA, 2006). The average annual temperature is 25°C, ranging from 11 to 38° C. It has a latitude of 15°'35'30" to the south and a longitude of 49°52'41" to the west and rainfall of 1600 mm per year (BARBOSA, 2006; COSTA-JÚNIOR; FIGUEIREDO, 2003).

Each area has different climates and latitudes, which can influence the results of the plant's natural chemical compounds. Abiotic and biotic factors influence the plant to use different strategies, such as defence strategies, which then oscillate its chemical compounds (SCHWAN-ESTRADA et al., 2008).

3.2 Collection of botanical material

The E. uniflora leaves were collected in the municipality of Itapuranga- GO, which has a latitude of 15º '35'30" to the south and a longitude of 49°52'41" to the west and a rainfall of 1600 mm per year (BARBOSA, 2006; COSTA- JÚNIOR; FIGUEIREDO, 2003). In Rua 24 n 136 vila Marilda sector xixazão, in the months of February and September 2017 as shown in figure 2.

Figure 3: Illustration of leaves collected from Eugenia uniflora species.
Source: Author, 2017.

The specimen under study was dried at the Experimental Biology Laboratory (LaBioex) of the State University of Goiás, Itapuranga Campus.

According to Pastorini, Bacarin and Abreu (2002) and Marure Sodek (1995), this process significantly reduces the plant's liquid content. This loss facilitates the preparation of plant extracts, which are easier to handle, transport and store, as well as favouring the maintenance of chemical, microbiological and pharmacological stability.

After drying, they were contained in plastic packaging and stored in a safe place to avoid insect attacks or any contamination that could influence the result. To prepare the extract and essential oil. The exsiccata of the plant under study

will be characterised and deposited in the Botanical Herbarium (HERBOT). After crushing the species, see figure 3 below.

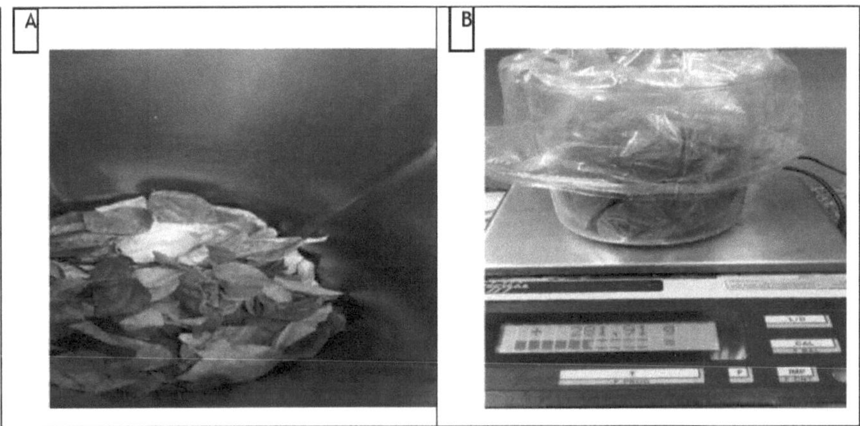

Figure 4: (A) Illustration of the leaves for shredding, (B) after shredding Source: Author, 2017.

3.3 Phytochemical prospecting

The aim of phytochemical research is to find out about the chemical constituents of plant species or to assess their presence in them. When there are no chemical studies available on the species of interest, phytochemical analysis can identify the relevant groups of secondary metabolites (SIMÕES, 2001). Such as flavonoids (SIMÕES, 2004), saponins (MATTOS, 1997) and coumarins Miranda et al (2013).

3. 4Essential oil

The method used to extract the essential oil was vapour drag, where each treatment was subjected to hydrodistillation in a Clevenger-type apparatus adapted to a round-bottomed flask for 90 minutes (OLIVEIRA et al., 2012).

The essential oil obtained was dried with $Na^2 SO^4$ and its yield was calculated using the initial weight of the bottle without the essential oil and the final weight of the bottle with the essential oil (ADAMS, 2007).

It was then packed in an impurity-free, hermetically sealed container and stored at low temperatures for later identification of the compounds present.

3.5 Identification of phytochemicals present in the oil

The essential oil obtained was analysed by gas chromatography coupled with mass spectrometry (GC/MS) using a SHIMADZU QP2010 Plus apparatus. Using a fused silica capillary column (CBP - 5; 30m x 0.25mm x 0.25μm), with a flow of 1mL/min of helium as the carrier gas, heating at a programmed temperature and an ionisation energy of 70 eV. With an injection volume of 1μL of each sample diluted in CH2Cl2 in a ratio of 1:5, after extracting the oil the calculation was made in a table to find out the data for the search for metabolites IR = 100 . N [(tx - tn-1)/(tn - tn-1)] + 100 . Cn-1 N = Cn - Cn-1 ADAMS, 2007).

3.6 In silico test to analyse pharmacokinetic properties

After identifying the compounds present in the essential oil, those that showed the highest percentages in the chromatogram of more than 5% of the oil's total constitution were selected for the analysis of pharmacokinetic properties.

The in silico analysis was carried out using two online programmes: MetaPrint2DReact, developed by the University of Cambridge available at: http://www- metaprint2d.ch.cam.ac.uk/, to predict the metabolism of the majority compounds; and Molsoft Drug-Linikess developed by Ruben Abagyan available at: https://www.molsoft.com/index.html, used to plot the pharmacokinetic profile of the majority compounds and the respective metabolites formed from them.

The data obtained using the Molsof Drug-Likiness software was used to analyse Lipinski's Five Rules, which are a set of empirical conditions that establish parameters for predicting the absorption profile and permeability of possible drugs. The evaluation criteria are: (I) number of hydrogen bond acceptor groups (nALH), which must be less than or equal to 10; (II) number of hydrogen bond donor groups (nDLH) less than or equal to 5; (III) molecular mass.

(MM) of no more than 500 g/mol; (IV) octanol-water partition coefficient (LogP) of less than or equal to 5; (V) polar surface area (PSA) of less than or equal to 140 Â2. Molecules that violate more than one of these rules may have bioavailability problems.

CHAPTER 4

RESULTS AND DISCUSSION

4.1 Analysing the phytochemical profile

The phytochemical analysis detected the presence of various secondary metabolites, including flavonoids, phenolic compounds, coumarins, digitalis and saponins. The secondary metabolites found in the leaves of E. uniflora are in agreement with the literature (AMORIM, 2009). As can be seen in Table 3 below:

Table 3: Class of secondary metabolites observed in E. uniflora leaves.

Secondary metabolites	Positive	Negative
Flavonoids	X	
Phenolic compound	X	
Coumarins	X	
Digitalics	X	
Saponin	X	

Source: author, 2017.

Flavonoids are a class of natural compounds of considerable scientific and therapeutic interest. According to Nijveldt (2008) "Flavonoids are a class of phenolic compounds that differ in their chemical structure (pg.1) Schmeda-hirschmann (1995)

also mentioned that the main flavonoids found in Eugenia uniflora leaves are quercetin and myricitrin. Phenols are organic compounds that have one or more hydroxyls (OH) attached directly to the aromatic ring. They have antibacterial and fungicidal activity and are widely used in the cleaning products industry for these properties. (ROCHA J., 20015).

Domingo; López-brea, (2003, p. 387) points out that the coumarin group (from the phenolic compound family) has attracted a great deal of biotechnological interest due to the different bioactivities attributed to some of its members, such as anti-inflammatory, antithrombotic and vasodilator properties. he points out that coumarin is one of the aromatic substances that can be used to flavour butter, tobacco, numerous medicines and perfumes (ROJAHN et a.l 1956).

Digitalis are cardioactive compounds that can inhibit the Na/K ATPase pump, resulting in an increase in the intracellular concentration of calcium and an increase in the intensity of the interaction of the actin and myosin filaments of the cardiac sarcomere, highly effective in the treatment of Congestive Heart Failure. They are highly absorbed orally, metabolised by the liver and excreted in the bile (CAUFFIELD JS, 2008).

Saponins have therapeutic activities such as digestive and diuretic, and are of great importance to the pharmaceutical industry (JENTSCH et al., 1961). Saponins are substances derived from the secondary metabolism of plants, mainly related to the defence system. They are found in tissues that are more vulnerable to fungal, bacterial or predatory insect attack (WINA et al., 2005).

4.2Analysing   the composition of essential oil from Eugenia uniflora L. leaves

The use of essential oils dates back to antiquity, when balsam, aromatic herbs and resins were used to embalm corpses. As time went by, they learnt more about the properties of these essential oils and began to be researched and used as a form of treatment for certain diseases. When they discovered secondary metabolites and complex chemical compounds, they realised the advantages of essential oils in the pharmaceutical industry (SOLÓRZANO-SANTOS; MIRANDANOVALES,2012).

Wolffenbuttel, (2007) quoted "Chemically speaking, essential oil is a natural oil, with a distinctive odour, secreted by the glands of aromatic plants, obtained by a physical process and with a chemical structure formed by carbon, hydrogen and oxygen, giving rise to a complex mixture of substances, which can reach several hundred of them, with a predominance of one to three substances that characterise the plant species in question. These substances have different structures, such as carboxylic acids, alcohols, aldehydes, ketones, esters, phenols and hydrocarbons, among others, each with its own aromatic characteristic and biochemical action."

The essential oil of Eugenia uniflora identified 20 chemical compounds in the leaf collected in February and 10 chemical compounds in the leaves collected in September using gas chromatography, as shown in figure 4 below:

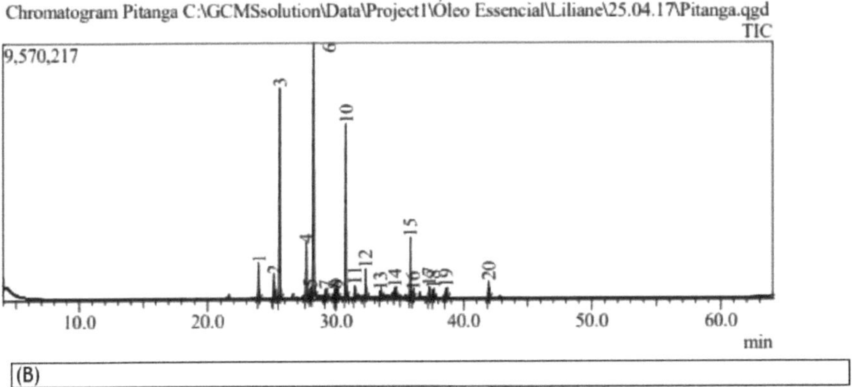

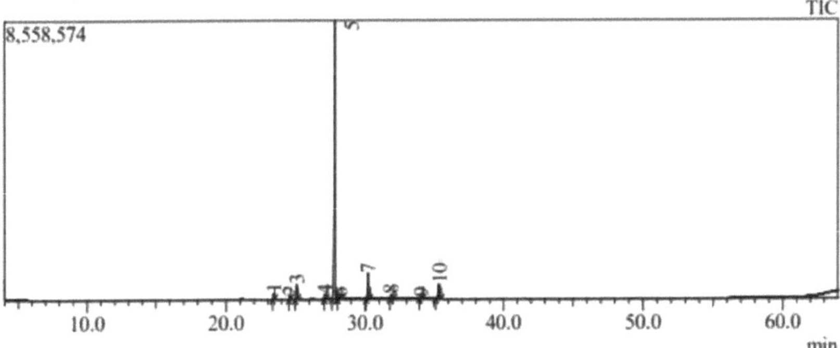

FIGURE 5: Chromatogram of the essential oil from E. uniflora leaves. (A) the chromatogram for February and (B) the chromatogram for September.

The GC-MS analysis identified 20 compounds in the February collection and only 10 compounds in the September collection, showing greater chemodiversity in the essential oil collected in February. However, the main compound found, curzerene, was obtained in a higher concentration, which could facilitate the isolation of the compound and a higher yield. The factors that may have influenced the chemodiversity and concentration of the metabolites are biotic and abiotic. In February the season was rainy, with high humidity and the leaves were budding, while in September the weather was dry and the low humidity left the leaves dehydrated, which may have influenced the concentrations of the results.Among the compounds obtained from the essential oil, the 5 most abundant are: curzerene, γ elemene, δ-germacreme, β-germacrame and germacrone as shown in Table 4.

Table 4: Chemical compound analysis

	Time of retention	Compound	January	August
			%	%
1	23.939	β Element	2.99	1.40
2	25.160	E caryophyllene	2.60	
3	**25.635**	**γ Elemeno**	**21.21**	**3.54**
4	**27.688**	**δ Germacrene**	**5.09**	1.08
5	27.998	β- Selinene	0.64	0.53
6	**28.290**	**Curzerene**	**31.39**	**78.12**
7	29.202	Δ Amarphene	0.78	
8	29.908	α- Cadinene	0.90	
9	30.092	Dauca-4 7 diene<trans	0.88	
10	**30.786**	**β-Germacrene**	**17.52**	**6.56**
11	31.460	Spathulenol	1.00	
12	32.311	Cis p-elemene	2.57	1.45
13	33.442	(Cis) Cadin -4-in-7-al	0.48	
14	34.613	Atractilone	1.11	0.69
15	**35.829**	Germacrone	**5.79**	**5.90**
16	36.019	Eudesm-7(11)-en-4-al	0.54	
17	37.285	Mint suffide	1.04	
18	37.619	Elemenone	0.79	1.45
19	38.619	Hydroxy Muurolene <14	1.19	
20	41.956	hydroxy α>	1.51	

The essential oil of E. uniflora leaves from Nigeria showed the presence of the most abundant components such as caryophyllene (5.7%), furanediene (24%), β-germacrene (5.8%), seline-1,3,7(11)-trien-8-one (17%) and oxidoseline-1,3,7(11)-trien-8-one (14%). Morais et al. (1996) isolated and identified the components of the essential oil of Eugenia uniflora L. leaves, harvested in the north-east of Brazil, with equivalent major components seline-1,3,5(11)-trien-8-one and oxidoseline-1,3,7(11)-trien-8-one.

4.3 Analysis of the metabolism of the major compounds in the essential oil of Eugenia uniflora leaves .

The processes by which drugs are altered by biochemical reactions in the body are collectively referred to as drug metabolism or biotransformation. In this study, the probabilities of reactions that the main compounds in the essential oil of E. uniflora leaves can carry out after biotransformation in the human body were evaluated. The compounds evaluated were: curzerene, γ-elemene, β-germacrene, germacrone and δ-germacrene. We observed that the main metabolic reactions were: hydroxylation, oxidation, epoxidation, hydration, reduction, glutathionisation and cysteamination. We observed the possible production of 3 metabolites for curzerene, β-germacrene, δ-germacrene and germacrone and 2 for γ-elemene, as shown in figure 6.

These biotransformation reactions include phase 1 (catabolic) reactions such as oxidation and hydroxylation, and phase 2 (anabolic) metabolism reactions such as glutathionisation.

Oxidative reactions in humans are most often carried out by hepatic cytochrome P450 oxidases, most of which exhibit broad substrate specificity. This is partly due to the activated oxygen in the complex, which is a powerful oxidising agent capable of reacting with a variety of compounds (RANG, et al. 2004).

Metabolic reactions are most often carried out by the hepatic system and are designed to make compounds more water-soluble in order to facilitate their elimination by the renal system. However, these enzymatic reactions can produce inactive, active and toxic compounds, as well as functionalising inactive compounds by inserting pharmacologically active groups, which occurs in the case of prodrugs (RANG, et al. 2004).

Therefore, the study of metabolism is important for assessing the behaviour of these compounds in the human body, evaluating their metabolic pathways and it can also be used as a way of producing new molecules, helping with chemodiversity, such as the main majority compounds found in the essential oil shown below:

Figure 6: Illustration of the main reactions that can occur with the major compounds found in the essential oil of E. uniflora leaves during human metabolism.

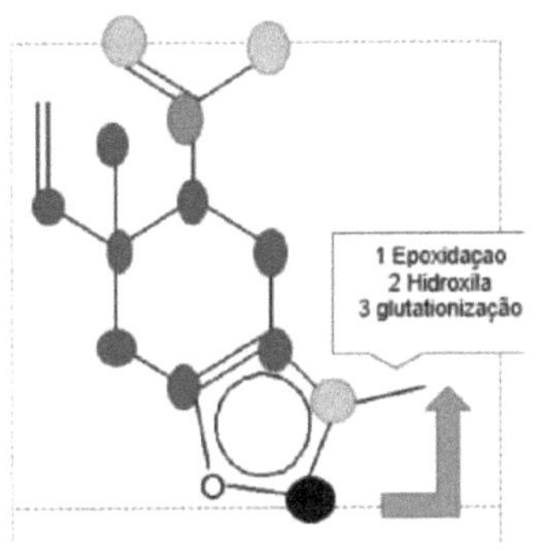

Curzerene 1(31.39%) 2(78.12%)

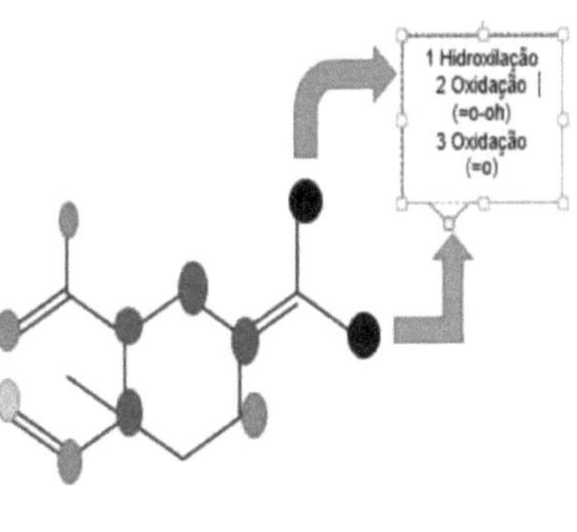

Yelemene 1(21.21%) 2(3.54%)

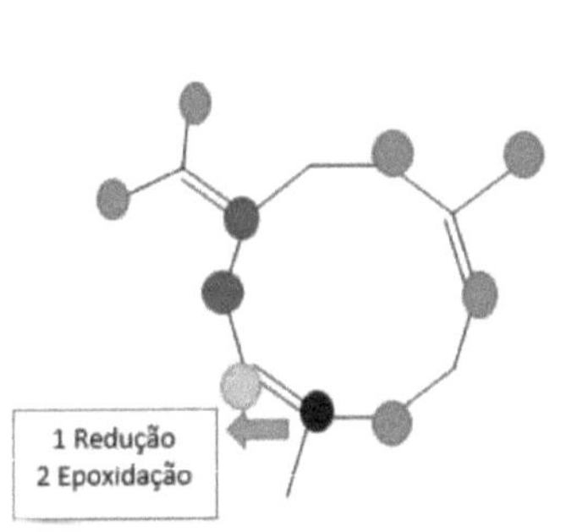

B germacrene 1(17.52%) 2(6.56%)

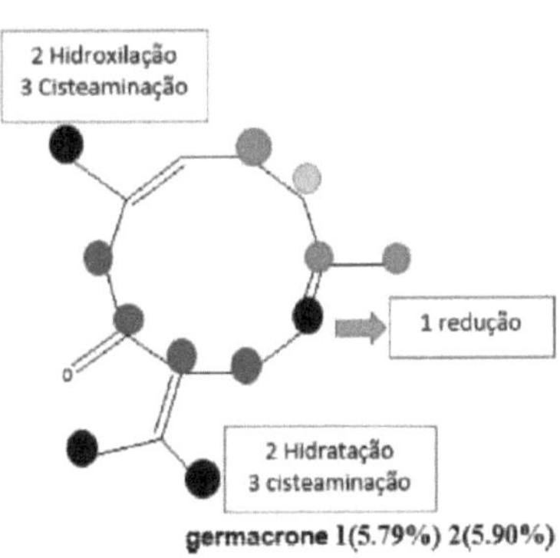

germacrone 1(5.79%) 2(5.90%)

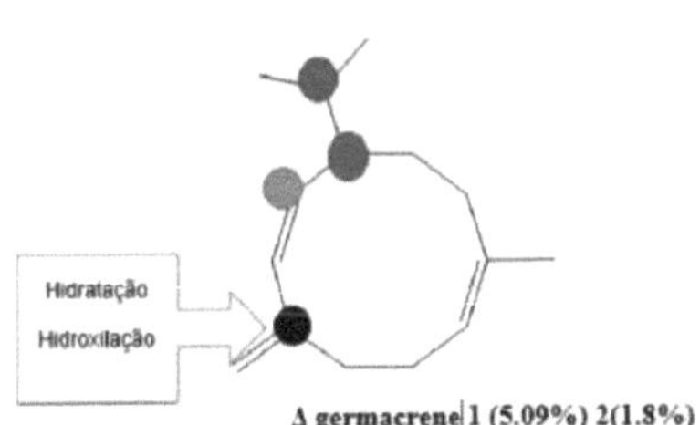

Δ germacrene 1 (5.09%) 2(1.8%)

0,66 <= NOR <= 1,00   0,33 <= NOR <0,66 0,15 <= NOR <0,33 0,00 <= NOR <0,15

Evaluating the majority compounds obtained from the essential oil of E. uniflora, we observed that Curzerene (31.38%; 78.12%) has great biological potential and is described for its moisturising, antimicrobial and anti-irritant properties, making it highly appreciated in the cosmetics industry (ROUSSIS, V; 2000).

The compound Y-elemene (21.21%; 3.54%) has demonstrated in vitro anti-proliferative effects against some types of cancer cells, indicating the possibility of its use in chemotherapy (PENG X, 2006).The third most abundant compound, β-germacrone (17.52%; 6.56%) has antimicrobial and insecticidal properties (FLAMINI G, CIONI PL, MORELLII, 2005). Germacrone has cytotoxic effects, inhibiting the proliferation of Bel-7402, HepG2, A549 and HeLa cells (LIANBAO YE, 2016).

The last major compound studied, δ-germacrene (5.09%; 1.8%), has antifungal properties (DEKRAKER et al., 1998).

The metabolic reactions obtained showed that most of them add polar compounds to the structure of the molecules, making them more soluble in water compared to the starting compound. In addition, we can mention that reactions such as the epoxidation observed in curzerene and β-germacream form compounds that are generally toxic.

4.4 Analysis of kinetic characteristics

Kinetics is an area of study that examines the forces that cause movement (HAMILL, 1999). Combining the concept of kinetics with the study of drugs, we have the line of study called pharmacokinetics. This study aims to analyse the behaviour of drugs, from their absorption in the human body to their distribution, metabolism, excretion and toxicity (ROMERO; ROMERO, 2014).

The permeability parameter described by Lipinski's rules proposes that compounds can present a maximum of one violation. According to this line of evaluation, γ-elemene, Δ-germacrene, B-germacrene, germacrone and their respective metabolites fit the bill as viable because they show good absorption and oral permeability. As shown in Table 5:

Table 5: kinetic characteristics through violation of lipinks.

Component	MW	HBA	HBD	LogP	LogS	Violation of Linpink's
Curzerene	220.18	1	0	4.30	-3.01	0
Metabolite 1	234.16	2	0	4.03	-2.73	0
Metabolite 2	232.15	2	1	4.07	-3.92	0
Metabolite 3	**521.31**	8	**8**	1.56	-4.59	2
Y-elemene	204.19	0	0	**5.69**	-3.27	1
Metabolite 1	220.18	1	1	4.71	-2.47	0
Metabolite 2	234.16	2	1	4.52	-2.22	0
Metabolite 3	218.17	1	0	4.38	-2.85	0
Δ germacrene	204.19	0	0	4.81	-3.77	0
Metabolite 1	222.20	1	1	3.89	-3.13	0
Metabolite 2	220.18	1	1	**5.33**	-3.23	1
Β germacrene	206.20	0	0	**5.68**	-4.81	1
Metabolite 1	222.20	1	1	4.45	-3.30	0
Metabolite 2	204.19	0	0	**5.14**	-4.42	1
Metabolite 3	218.20	0	0	**5.63**	-5.63	1
Germacrone	218.17	1	0	**5.15**	-3.89	1
Metabolite 1	220.18	1	0	4.80	-4.11	0
Metabolite 2	234.16	2	1	4.17	-2.92	0
Metabolite 3	293.18	3	3	4.17	-4.15	0

The data obtained by analysing the pharmacokinetic properties show that curzerene, the compound that was found in the greatest quantity in the essential oil, is expected to form 3 metabolites. Of these, only metabolite number 3 violated Lipinski's rules, both in terms of molecular mass, which required a maximum of 500 g/mol, and nDLH, which had to be equal to or less than five (Table 05). Reading this data shows that although curzerene has good permeability, one of the metabolites formed from it may have bioavailability problems.

The five main majority compounds and their metabolites were

evaluated according to Lipinks' rule. Those that violated the rule were curzere metabolite 3, which violated the MW being hydrosolubility and the electron donor HBD, γ elemene, metabolite 2 of δ germacrene, B germacrene and respectiveis metabolite 2 and 3 and germacrone also violated the LogP being a partition coefficient in water and octanol.

Pharmacokinetic modelling makes it possible to predict the movement of chemical compounds in the human body and can assess properties such as absorption, distribution, metabolism and excretion. To do this, mathematical calculations can assess whether a compound has chemical characteristics that promote good absorption, such as permeability, oil/water partition coefficient, particle size, pH and others (GALLO NETO, 2012).

The importance of studying these pharmacokinetic parameters lies in the fact that they affect the concentration of the drug in the blood and at the target site and thus its pharmacological effect. A drug with low permeability and a low partition coefficient is generally poorly absorbed and therefore may not reach the target site in sufficient concentration, thus not achieving a therapeutic effect (PENILDON, 2012). Drugs with a high oil-water partition coefficient generally have a high binding capacity with plasma proteins and the ability to bioaccumulate, which can also influence the active concentration of the drug and its effect (HAMILL, 1999).

Evaluating the kinetic properties means analysing whether the compound has a balance between the parameters so that it is absorbed, distributed, metabolised and excreted, together guaranteeing the biological effect with adequate intensity and time (GALLO NETO, 2012).

CHAPTER 5

FINAL CONSIDERATIONS

The essential oil from the leaves of E. uniflora L. showed a fairly wide chemidiversity in January, but in August the number of compounds decreased substantially. These data show how biotic and abiotic effects can influence the composition of the essential oil. More research is therefore needed in this area to find out which environmental characteristics provide the highest yields of essential oil.

In general, all the major compounds are in accordance with Lipinski's Five Rules. These data, together with the bibliographical survey on therapeutic properties, serve as an incentive for further research into the potential use of each compound. Further screening could be carried out on the isolated compounds to see if any of them are toxic, as well as other in silico, in vitro or in vivo tests.

CHAPTER 6

REFERENCES

Amorim A.C.L., C.K.F. Lima, A.M.C. Hovell, A.L.P. Miranda, C.M. Rezende, Antinociceptive and hypothermic evaluation of the leaf essential oil and isolated terpenoids from Eugenia uniflora L. (Brazilian Pitanga), Phytomedicine 16 (2009) 923-928.

Almeida, E.C., Karnikowski, M.G.O., Foleto, R.,Baldisserotto, B. Analysis of antidiarrhoeic effect of plants used in popupal medicine. Rev. Saúde Pública, 29:428- 433, 1995.

Alonso, J. R. Tratado de Fitomedicina. Clinical and Pharmacological Bases. Buenos Aires: Isis Ediciones S.R.L., 1998.

ANGIOSPERM PHYLOGENY GROUP III, Engelska. Ingâr i: Botanical Journal of the Linnean Society. ; 161:2, s. 105-121 (2009).

Araujo, G. SALVO EM PLANTAS <www.remedio-caseiro.com/goiabeira- benefits-and-properties/> Accessed 11 Jan. 2017

Auricchio, M. T.; Bacchi, E. M. Leaves of Eugenia uniflora L. (pitanga): pharmacobotanical, chemical and pharmacological properties. Rev. Inst. Adolfo Lutz, São Paulo, v. 1, p. 55-61, 2003.

Cairns J. The origin of human cancers. Nature 289,353-7, 1981.

Cauffield JS, Gums JG, Grauer K. The serum digoxin concentration: ten questions to ask. Am Fam Physician. 1997;56:495-503;509-10.

The genus Eugenia: from chemistry to pharmacology. Veloso, Juvenal henrique. 2 Calixto, J. B.; Siqueira júnior, J. M. Development of medicines in Brazil: challenges. Gazeta Médica da Bahia. Bahia, v.78, supplement 1, p.98- 106, 2008.

Consolini, A. E. ; BALDINI, O. A. N. ; AMAT, A. G. Pharmacological basis for the empirical use of Eugenia uniflora L. (Myrtaceae) as antihypertensive. Journal of Ethnopharmacol, Limeric. v. 66, p. 33-39, 1999.

ConsolinI, A. E.; SARUBBIO, M. G. Pharmacological effects of Eugenia uniflora (Myrtaceae) aqueous crude, extract on rat' heart. Journal of Ethnopharmacology. v. 81, p. 57-63, 2002.

Cechinel Filho, V.; YUNES, R. A. Strategies for obtaining pharmacologically active compounds from medicinal plants. Concepts on structural modification to optimise activity. Química Nova, v. 21, n. 1, p. 99-105, 1998.

Danchin A, Medigue C, Gascuel O, Soldano H, Henaut A. From data banks to data bases. Res Microbiol. 1991 Sep-Oct;142(7-8):913-6

deKraker, J.-W., Franssen, M.C.R., de Groot, A., Konig, W.A. and Bouwmeester, H.J. (1998) (+)-Germacrene A biosynthesis. The committed step in the bitter sesquiterpene lactones in chicory. Plant. Phys. , 117, 1381-1392.

Domingo, D.; LÓPEZ-BREA, M. Plantas com acción antimicrobiana. Revista Espanola de Quimioterapia, v. 16, n. 4, p. 385-393, 2003

Hussar A. Daniel Merck Sharp & Dohme Corp., subsidiary of Merck & Co., Inc., Kenilworth, NJ, USA, 2017.

Fiuza, tatianas ., et al. "Pharmacognostic characterisation of the leaves of Eugenia uniflora L. (Myrtaceae)." Revista Eletrônica de Farmácia 5.2 (2008).

Fiuza, t.s.; Silva, p.c.; Paula, jr.; Tresvenzol, l.m.f.; Souto, m.e.d.; Sabóia-morais,s.m.t. Tissue and cellular analysis of the gills of Oreochromis niloticus L. treated with crude ethanolic extract and fractions of pitanga leaves (Eugenia uniflora L.) - Myrtaceae. Revista Brasileira de Plantas Medicinais: Botucatu, v.13, n.4, p.389-395, 2011.

Flamini G, Cioni PL, Morelli I (2005). "Composition of the essential oils and in vivo emission of volatiles of four Lamium species from Italy: L. purpureum, L. hybridum, L.

bifidum and L. amplexicaule". Food Chemistry. 91 (1): 63-68

Guido, R. V. C.; Andricopulo, A. D.; Oliva, G. Drug planning, biotechnology and medicinal chemistry: applications in infectious diseases.

Revista Estudos Avançados. São Paulo, v.24, n.70, p.81-98, 2010.

KAPLAN, W. et al. Priority Medicines for Europe and the World 2013 Update. 9 Jul 2013. 2013

Lee, M-H.; Chiou, J-F.; Yen, K-Y.; Yang, L-L. EBV DNA polymerase inhibition of tannins from Eugenia uniflora,. Cancer Letters, Amsterdam. v. 154, p. 131-136, 2000.

Lewinsonhn, T. M.; Prado, P. I.; Biodiversidade Brasileira - Síntese do

Estado Atual do Conhecimento, 1ª . ed., Ed. Pinsky: São Paulo, 2002, chap. 1, p. 17-25.

Lima, V. L. A. G.; Mélo, E. A.; Lima, D. E. S. Total phenolics and carotenoids in pitanga. Scientia Agricola, v. 59, n. 3, p. 447-450, 2002
Lombardino, J. G.; Lowe III, J. A.; Nat. Rev. Drug Discov. 2004, 3, 853.

Maia, J.G.S., Andrade, M.H.L., Da Silva, M.H.L., Zoghbi, M.G.B. A new chemotype of Eugenia uniflora L. from north Brazil. J. Essent. Oil Res., 11:727-729, 1999.

Maia, J.G.S., Zoghbi, M.G.B., Luz, A.I.R. Essential oil of Eugenia punicifolia (HBK) DC. J. Essent. Oil Res., 9:337- 338, 1997.

Malaman, F.S.; Moraes,L.A.B.; West, C.; Ferreira, N.J.; Oliveira, A.L. Supercritical fluid extracts from the Brazilian cherry (Eugenia uniflora L.): Relationship between the extracted compounds and the characteristic flavor intensity of the fruit. Food Chemistry 124: 85-92, 2011.

Mendes A. Factorial design andresponse surface optimisation of crude violacein for
Chromobacterium violaceum production Biotechnology Letters, 2006)

Morais, S.M., Craveiro, A.A., Machado, M.I.L., Alencar, J.W., Matos, J.A. Volatiles constituents of Eugenia uniflora leaf oil from northeastern Brazil. J. Essent. Oil Res., 8:449-

Nath, B. S.; Kumar, R. P. S. Ecotoxicol. Environ. SAF. 1999, Ottensmeier, 2001; Paes et al. 2002; Kumar et al. 2008
Mendes AS. , Carvalho JE, Duarte MC, Durán N, Bruns RE. Factorial design andresponse surface optimization of crude violacein for

Chromobacterium violaceum production Biotechnology Letters.

Palmeira filho, P. L.; Pan, S. S. K. Cadeia farmacêutica no Brasil: avaliação preliminar e perspectivas. BNDES Sectorial. Rio de Janeiro, v.1, n.18, p.3-22, 2003.

Peng X, Zhao Y, Liang X, Wu L, Cui S, Guo A, Wang W (Feb 2006). "Assessing the quality of RCTs on the effect of beta-elemene, one ingredient of a Chinese herb, against malignant tumours". Contemp Clin Trials. 27 (1): 70-82.

Rang, H P, et al. Pharmacology. [P L Voeux and A J M S Moreira. 5ª Edition. Rio de Janeiro : Elsevier, 2004.

Schwan-estrada, K. R. F.; Stangarlin, J. R.; PascholatI, S. F.
Biochemical mechanisms of plant defence. In: PASCHOLATI, S.F.; LEITE, B.;
Stangarlin, J.R.; CIA, P. (Ed.). Plant Pathogen Interaction - Physiology, Biochemistry and Molecular Biology. Piracicaba: FEALQ, 2008. p.227-248.

Silva, A. R. Development of new analytical methodologies for monitoring PPCPs in real matrices. 2010. 290 f. Thesis (PhD in Chemistry) - Faculty of Sciences, University of Lisbon, Lisbon.

Solórzano-santos, F.; Miranda-novales, M.G. Essential oils from aromatic herbs as antimicrobial agents.Current Opinion in Biotechnology, v. 23, n. 2, p. 136 141,2012.

Tozer, T. M.; Rowland, M. Introduction to pharmacokinetics and pharmacodynamics: the quantitative bases of pharmacological therapy. Porto Alegre: Artmed, 2009. 336 p.

Zhong, Z.; Chen, X.; Tan, W.; Xu,Z.; Zhou, K.; Wu, T.; Cui, L. And Wang, Y. Eur.J.Pharmacol. 2011, 667, 50.

Wazlawik, E.; Silva, M. A. da.; Peters, R. R.; Correia, J. F.; Farias, M. R.; Calixto, J. B.; Ribeiro-do-valle, R. M. Analysis of the role of nitric oxide in the relaxant effect of the crude extract and fractions from Eugenia uniflora in the rat thoracic aorta. Journal of Pharmacy and Pharmacology, v. 49, n. 4, p. 433-437, 1997.

Wolffenbuttel, Adriana Nunes. Essential oils. CRQ-V Newsletter, year XI, no. 105, pages 06 and 07 November/December/2007.

Wyerstahl, P., Marschall- Wyerstahl, H., Christiansen, C., Oguntimein, B.O., Adeoye, A.O. Volatile constituents of Eugenia uniflora L.leaf oil. Planta Med.,6:546-549, 1988.

I want morebooks!

Buy your books fast and straightforward online - at one of world's fastest growing online book stores! Environmentally sound due to Print-on-Demand technologies.

Buy your books online at
www.morebooks.shop

Kaufen Sie Ihre Bücher schnell und unkompliziert online – auf einer der am schnellsten wachsenden Buchhandelsplattformen weltweit! Dank Print-On-Demand umwelt- und ressourcenschonend produziert.

Bücher schneller online kaufen
www.morebooks.shop

Printed by Books on Demand GmbH, Norderstedt / Germany